Soumahoro Soualiho

Local food production

Soumahoro Soualiho

Local food production

and supply of the Boundiali urban market (Côte d'Ivoire)

ScienciaScripts

Imprint
Any brand names and product names mentioned in this book are subject to trademark, brand or patent protection and are trademarks or registered trademarks of their respective holders. The use of brand names, product names, common names, trade names, product descriptions etc. even without a particular marking in this work is in no way to be construed to mean that such names may be regarded as unrestricted in respect of trademark and brand protection legislation and could thus be used by anyone.

Cover image: www.ingimage.com

This book is a translation from the original published under ISBN 978-620-6-72417-9.

Publisher:
Sciencia Scripts
is a trademark of
Dodo Books Indian Ocean Ltd. and OmniScriptum S.R.L publishing group

120 High Road, East Finchley, London, N2 9ED, United Kingdom
Str. Armeneasca 28/1, office 1, Chisinau MD-2012, Republic of Moldova, Europe
Printed at: see last page
ISBN: 978-620-8-15828-6

LOCAL FOOD PRODUCTION AND SUPPLY TO THE URBAN MARKET IN BOUNDIALI (CÔTE D'IVOIRE)

LOCAL FOOD PRODUCTION AND SUPPLY TO THE URBAN MARKET IN BOUNDIALI (IVORY COAST)

SOUALIHO SOUMAHORO

PELEFORO GON COULIBALY UNIVERSITY KORHOGO

SSOUALIHO20@GMAIL.COM

SUMMARY

Perennial agriculture and mining are the main economic activities in Boundiali. Media coverage of the cotton bonanza obscures the policy of self-sufficiency in local food crops. The marginalisation of food-producing agriculture is therefore back in the spotlight, creating a strong demand for food in the study area. However, a socio-cultural organisation has been set up by rural populations to supply the Boundiali urban market.The aim of this study is to analyse the positive impact of local food production in supplying the Boundiali market. The methodological approach was based on documentary research and a field survey using a questionnaire and an interview guide. The results show that maize (37.85%), groundnuts (37.38%), rice (22.33%) and yams (1.87%) are the dominant foodstuffs offered by local producers on the Boundiali market. Chicken (36.19%), goat (8.25%), mutton (23.92%), beef (27.20%), pork (1.27%) and guinea fowl (3.17%) are the main food animal products in the study block. In addition, long-distance transport supplies the Boundiali market with food products. Finally, an average contribution of €2,869.34 per year benefits the local community through the levying of taxes on the urban market. However, the marketing of food products is leading to a deterioration in the environment of the Boundiali market.

Key words : Boundiali, urban supply, food crops, market.

INTRODUCTION

Côte d'Ivoire's economy is based on industrial agriculture. Since colonisation, food production has had a low profile. In fact, the euphoria of the country's years of economic prosperity masked the problem of self-sufficiency. Decades later, the urbanisation movement, with its attendant problems, is raising collective awareness. Côte d'Ivoire was characterised by accelerated urbanisation in the post-independence years. Anne Marie COTTEN (1968: 225) has estimated that among the French-speaking countries of West Africa, Côte d'Ivoire is the one where the urban phenomenon is appearing most spectacularly. Estimated at 6,709,600 inhabitants in 1975 (DCGTX, 1975), the number of inhabitants had risen to 15,366,672 by 1998 (DCGTX, 1998). The country's population will increase from 22,671,331 in 2014 to 29,389,150 in 2021 (INS-RGPH, 1998, 2014, 2021), representing an annual growth rate of around 2.9% for the period 2014-2021, with an urban population of around 15,152,232, or 53.9%, compared with 12,944,419 in rural areas, or 46.1%. It is one of the most urbanised countries in sub-Saharan Africa, with 43% of the total population living in urban areas (TAPE BIDI et al., 2018: 1). What is more, the number of urban dwellers reached cruising speed in 2013 at 53.26% (KOFFIE-BIKPO and ADAYE, 2014: 141-149). At the same time, the number of Ivorian cities has risen from 30 to 89 and to 127 today (INS-RGPH, 1988, 1998, 2014), with the city of Abidjan dominating (CIPD +30, 2023: 1). Like the Abidjan metropolis, the Bagoué town of Boundiali has a population of 6,5191, compared with 3,9962 since 2014, representing a growth rate of 24.00% (INS-RGPH, 2014; 2021). It covers an area of around 33 km^2 (Google Earth/satellite image, 2022). This growth in the number of inhabitants means that the area is expanding. The consequence is an increase in the number of people to feed. As a result, strong demographic growth is raising serious concerns (Patience MPANZU BALOMBA, et al., 2011: 1). The staggering growth in the population is quadrupling the need for food (Mendez del Villar P, Bauer JM, 2013, pp 4-9). Urban-rural relations are important for meeting people's needs (HOUNGBO, 2015: 13-14). In addition, cities were correctly supplied by the countryside (CHALEARD.J.L., 1996: 111-122). The endogenisation of supply is therefore central. However, the ability of the African countryside to feed the cities has been a subject of great controversy since the 1970s (MPANZU BALOMBA et al., 2011, op. cit.). In any case, the development of agriculture is necessary to control urban growth in Côte d'Ivoire (Jacqueline PELTRE-WURTZ and Benjamin STECK, 1991: 8). It provides food for both rural and urban populations (DUFUMIER. M, 1999: 547-560). To achieve this social objective, the Ivorian authorities set up development companies (SODEs): CIDT for cotton, SODESUCRE for sugar, SODEPRA for livestock, SODERIZ for rice and SODEFEL for vegetables between 1960 and 1980. The desire to achieve food self-sufficiency was clear-cut, as these state companies were endowed with substantial financial resources and structures for supervising farmers (KONAN et al., 2016: 6). They were firmly established in the Boundiali area. In other words, the savannahs of the north are ideal for food crops (YABILE, 1986: 40). Alongside these national tools, the International Fund for Agricultural Development (IFAD) has been funding cereal and market garden projects since 2017. Previously, these were the agricultural promotion and marketing support project (PROPACOM) and now the agricultural commodity chain development support programme (PADFA). The authorities' ambition with these projects is certainly to improve the processing and marketing of rice, but also to work towards an agricultural flowering, guaranteeing self-sufficiency. However, the environment in the Boundiali region has been marked by psychosis due to rumours of terrorist attacks in the villages. In the background, the cotton crisis of the 1980s, with its attendant decline in household purchasing power, deprived the State of the foreign currency it needed to subsidise food imports. Lastly, the corona virus (covid-19) pandemic severely slowed the flow of goods. This bleak picture is leading to shortages of essential products, especially on urban markets. The aim of this study is to analyse the impact of local food production in supplying the Boundiali market.

MATERIALS AND METHODS

I-Presentation of the study area and equipment

Boundiali is the capital of the Bagoué region. Located 100 km from Korhogo in the north-west, it straddles parallels 9°29 and 9°32 north latitude and 6°27 and 6°29 west longitude (Figure 1). It is characterised by a lack of major river systems and a dry climate with annual rainfall ranging from 64 to 86 days (UN-Habitat, 2012: 7).

Figure 1: Location of the study area

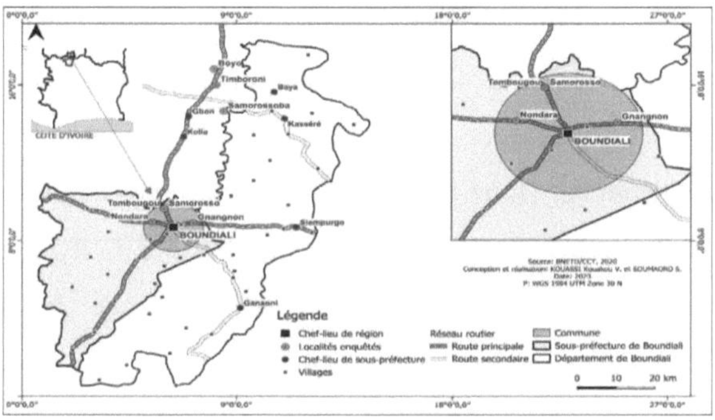

Source: Fieldwork, 2022-2023

II-Methodological approach

A rigorous scientific approach was adopted in order to achieve our objectives. Documentary research was carried out in libraries and documentation centres. The library of the Peleforo Gon Coulibaly University in Korhogo (UPGC), the Korhogo municipal library, the Regional Department of Agriculture, Rural Development and Food Production in Boundiali, the ANADER regional office in Korhogo and the ANADER branch office in Boundiali, the Centre National de Recherche Agronomique (CNRA) in Korhogo and the Office pour la

Commercialisation des Produits Vivriers (OCPV) in Abidjan. The Bureau National d'Etude Technique et de Développement (BNETD) and the Institut National de la Statistique (INS) in Korhogo were visited to consult cartographic documents. Internet search engines such as Google scholar were also used to collect secondary data. With regard to primary data collection, the field survey was conducted from November 2021 to February 2022. To make the research more effective, a questionnaire was used to cover the entire study block and to question the food producers. Interviews were also conducted with Boundiali town council technical services officers regarding the number of markets, the regularity of local food products and the collection of municipal taxes. The individuals and rural localities to be surveyed were selected as follows: the study localities were chosen within a 10 km radius of the town centre. All four (04) localities were surveyed. As for the selection of farm households to be interviewed, we used the reasoned choice technique, imposing criteria relating to the quality of the producer (food crop producer), the status of the farmer (at least one year), age (at least 25 years), nationality and the specificity of the food crops (maize, rice, groundnuts, millet, sorghum, beans, yams). A sample of The representative sample was drawn on a sampling basis from a sub-prefectural agricultural population of 1,550, from which 55 heads of household from the selected villages were deducted. The sample size was determined using D. Schwartz's probabilistic method (1995: 209), based on the assumptions made for identifying the target villages and individuals:

X= [(**za**^{2}) x **p** **(1-p)**] /d2; **X** being the sample size, **za** the 95% confidence level (standard value of 1.96), **d** the margin of error at 5% (standard value of 0.05), **q=1-p**, **p** = **n/N** with **p** the proportion of farm households concerned with food crops in the farm households of each village in the commune to be surveyed (n) in relation to the total number of farm households in the commune (N) in which the latter is located. By proceeding in this way for each village and dividing the result by 10%, we obtain the following numerical application for the village of

Gnagnon: $X= [(1.96)^2 \times 0.018 (1-0.018)] / (0.1)^2$ which is equal to how many six (6) heads of farming households. The results are shown in Table 1.

Table 1: Number of heads of farming households surveyed in the study area

Communes	Selected locations	Total number of heads of food crop farming households by gender, by locality withholding	Total number of heads of food crop farming households surveyed by locality
Boundiali	Gnagnon	02/26	00/06
	Nondara	07/138	02/31
	Samorosso	02/21	00/05
	Tombougou	03/48	00/11
Total	04	14/233	02/53

Blue=Female; Orange=Male

Source: National Statistics Institute (INS-RGPH, 2014)

Table 1 shows that the field survey covered 55 heads of agricultural households out of a total of 247. Thus, 2 female heads of household and 53 male heads of household out of a total of 14 female heads of household and 233 male heads of household were interviewed. The studies in the Boundiali market focused on data relating to the actors who carry out activities in this market. To carry out this work, a field survey was carried out during the first twenty days of June 2022. For greater rigour, a questionnaire was used as a roadmap for visiting the entire research framework by administering questions to traders. Given the impossibility of conducting an exhaustive survey given our resources, a sample was drawn up based on the postulates of identifying the target markets and people. The research framework covers one market, in this case a single market: the Boundiali market, which has 952 traders out of an estimated regional total of 3,848 (INS- RGPH; Mairie de Boundiali, 2014). The formula developed by DAKOURO Guissa Desmos Francis and KOULAÏ Armand (2015, p 68) has been used as a basic principle for the constitution of the sample, namely :

Mi= a (Ci/Ct) x100 where **Mi=Number of** traders to be surveyed per market; **Ci=Number of** traders present in the market; Ct=Total **number of** traders in the region's markets, **a=Constant** to enlarge. Example1: Number of traders to be surveyed in the Boundiali markets. Ci=950 traders, Ct=3848, a=0.5, Mi=0.5(950/3848) x 100=12, 34 approximately 12 people.

A total of 12 traders were interviewed in the Boundiali market out of a total of 950. Particular attention was paid to the food crops sector. A number of criteria were used to select the food traders in the research area. These included gender, nationality, level of education and number of years spent in the markets. Thus, at least one year of existence was retained. Statistical processing of the data was made possible by computer tools. Word version 2010 was used to draft the document, while Sphinx version 4.5 was used to create the questionnaire. Microsoft Excel version 2013 was then used to represent the tables and diagrams. QGIS version 3.12 was used to design and produce the maps, without calling into question the important role played by the Global Positioning System (GPS) in geo-referencing the position of the market squares used to sell food products, and the precise location of the rubbish dumps in the market's trading area. Finally, the chi-square test, or test of independence between two characteristics: a dependent qualitative variable and an independent qualitative variable, made a vital contribution. The calculation of this index is based on two hypotheses:

➢ **HO: so-called null hypothesis**, which assumes that the characteristics are independent.

In practical terms, this means that there is no link between the two variables;

➢ **H1: so-called alternative hypothesis**, which contradicts the previous hypothesis, since it is assumed that there are a link between the two variables. It is calculated by :

$$\mathbf{X^2 = (Observed\ workforce - Theoretical\ workforce)^2 / Theoretical\ workforce}$$

With X^2 , the chi-square value. If :

-Calculated chi-square < theoretical chi-square, then the two variables are independent;

-Calculated Chi-square > Theoretical Chi-square, then there is a link between the two variables. In addition, the Cramer test is used to measure the strength of the relationship between the two variables. This index is calculated by :

V= Square root (X2/n*[min (l, c) -1])

With n: sample size; min l: minimum of row or column (c) in the table; minimum of column in the table. Cramer's contingency coefficient is between 0 and 1. If :

-V=0, the two variables are independent;

-V is close to 0, so the link between the two variables is very weak;

-V is close to 0.25, so the link is weak;

-V is close to 0.5, so the link between the two variables is moderately strong;

- V is close to 0.75, so the link between the two variables is strong;

-V is close to 1, the link between the two variables is very strong.

RESULTS

I- Producer profile

I-1-Low presence of women in subsistence farming

Food crop farming in the study block is carried out by both men and women. However, women occupy a less honourable place in food production. They are obliged to perform other tasks. This argument is justified by examining Figure 2, which sheds light on the subordination of female agricultural labour to that of men. Figure 2 shows that, in general, men dominate agricultural activity in Gnagnon, Samorosso, Tombougou and Nondara. Only Nondara has 2 female heads of farming households out of 31, i.e. 6.45%, compared with 3.63% for all the heads of households surveyed in the 4 localities. In all, 53 out of 55 heads of household are male, or 96.36% of the heads of agricultural households surveyed. The domination of men to the detriment of women in agricultural activity in the rural areas of the study zone could be justified by the fact that women's role is confined to the home, where they prepare hearty meals for their husbands. Photo 1 shows farmers clearing land for food crops in Tombougou, 3 km from Boundiali. This study on gender in food crop farming by the research team could be a wake-up call to those in power, who need to redouble their efforts and vigilance to ensure that the policy of parity between male and female children is respected and applied.

Figure 2: Proportion of household heads surveyed by gender

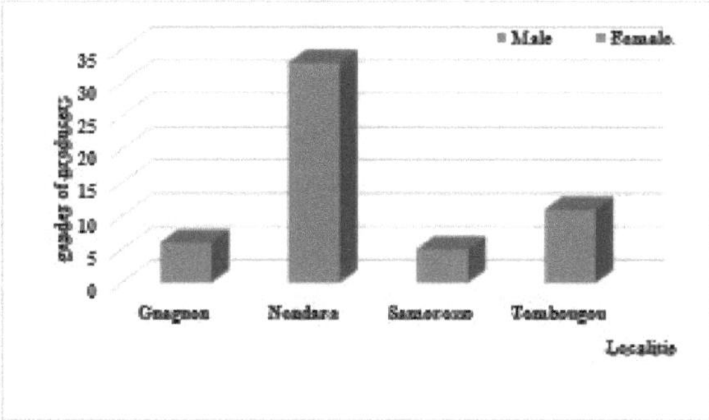

Source: Fieldwork, 2022-2023

Photo 1: Producers preparing a space for growing food crops in Samorosso

Source: SOUMAHORO photo, 2022

The level of education of food crop growers in the study area could also partly explain the nature of the cultivation methods and techniques used.

1-2-Low level of education among rural producers

Farmer-producers of food crops have a low level of education. It varies between no schooling, primary school and secondary school. Figure 3 shows the educational level of farmers according to the educational institution attended. Figure 3 shows that 65.45% of heads of farming households are illiterate, followed by primary school (20%) and secondary school (14.55%). Gnagnon has the lowest illiteracy rate, estimated at 5.45%, and Nondara the highest (34.55%). Nondara also has the highest rate of heads of household at primary (90.91%) and secondary (36.36%) levels. The high proportion of heads of agricultural households with no schooling among the heads of households surveyed is thought to be due to the maintenance of certain traditional constraints that are a hallmark of the people's cultural identity. senoufo. Thus, school education is not a priority for these people. It seems that the low level the quality of food production in the study block.

Figure 3: Heads of educated households by level of education

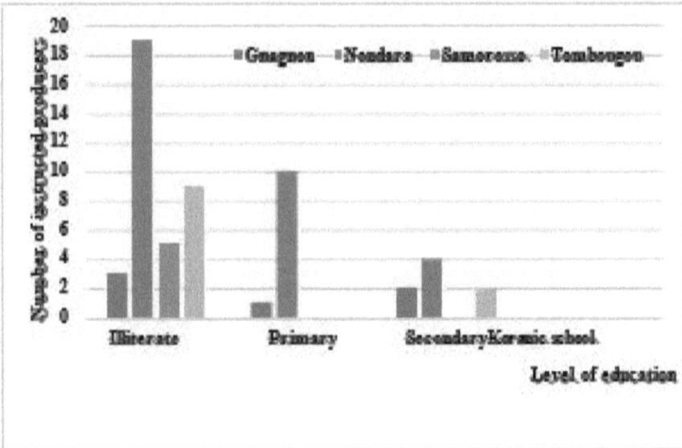

Source: Fieldwork, 2022-2023

I-3-1-Food production independent of producers' level of education

Food production is carried out by farmers, the majority of whom cannot read, write or do arithmetic. Table 2 highlights the link between farming activity and level of education.

Table 2: Correlation between age and food production

STAFFING TABLE OBSERVED					
Products	Rice	But	Peanut	Yam	TOTAL
Not enrolled	69	88	65	25	247
Educated	22	29	25	10	86
TOTAL	91	117	90	35	333
THEORETICAL STAFFING TABLE					
Products	Rice	But	Peanut	Yam	TOTAL
Not enrolled	67,50	86,78	66,76	25,96	247
Educated	23,50	30,22	23,24	9,04	86
TOTAL	91	117	90	35	333
CALCULATED CHI-SQUARE TABLE					
Products	Rice	But	Peanut	Yam	TOTAL
Not enrolled	0,03	0,02	0,05	0,04	0,13
Educated	0,10	0,05	0,13	0,10	0,38
TOTAL	0,13	0,07	0,18	0,14	0,51
Degree of freedom		(L-1) *(c-1)			
Degree of freedom		3			
Significance threshold			5%		
Theoretical chi-squares				7,82	

Source: Fieldwork, 2022-2023

Looking at table 2, the calculated Chi-square is lower than the theoretical Chi-square, i.e. 0.51<7.82. The analysis of this variable shows that there is no link between the level of education and the practice of subsistence food production. In addition to these factors, there are other factors that could support food production in the research environment.

I-3-Main d'œuvres adultes pour la pratique d'une agriculture vivrière

Agriculture is the main activity in the study area. It employs over 90% of the working population. The agricultural workforce in the research area is made up of the older population: adults, as defined in Table 3.

Table 3: Average age of growers by production area

Location	Average ages	Proportion of average ages (%)
Gnagnon	44	22,79
Nondara	42	21,76
Samorosso	58	30,05
Tombougou	49	25,39
TOTAL	193	100

Source: 2022-2023 field survey

Looking at Table 3, the rural localities in the research area have an average age of around 48.25 years. Clearly, Gnagnon has 6 heads of household with an average age of 44, i.e. 22.79% of the heads of household in the 4 localities surveyed; Nondara 42 years, i.e. 21.76%; Samorosso 58 years, i.e. 30.05%; Tombougou 49 years, i.e. 25.39%. Figure 4 shows the proportion of heads of food crop farming households in the research area by age group.

Figure 4: Proportion of age categories of food-producing farmers

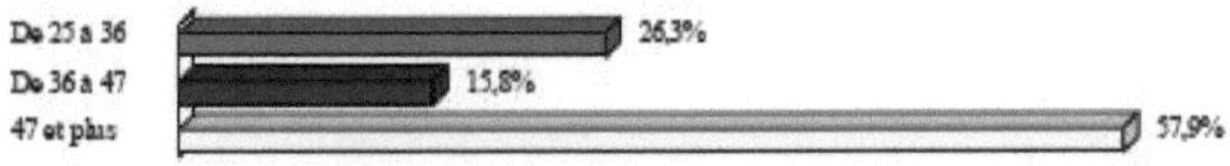

Source: Fieldwork, 2022-2023

Analysis of figure 4 shows that the heads of household are grouped into three (3) age categories: [25 to 36 years old], [36 to 47 years old] and [47 years and over]. 26.3% of heads of farming households in the 4 rural areas surveyed were aged between 25 and 36. 15.8% of heads of household are between 36 and 47. Finally, 57.9% are aged 47 and over. These graphs show that the heads of farming households producing food products in 47 and over are more numerous than those aged [25 to 36 [(26.3%) and [36 to 47 [(15.8%). Does age influence the practice of food production?

I-3-1-Food production independent of the age of producers

Men and women are the main workforce available for food production. Table 4 analyses the relationship between food crop production and the age of producers in the research block.

Table 4: Correlation between grower age and food production

STAFFING TABLE OBSERVED					
Products	Rice	But	Peanut	Yam	TOTAL
Adults	65	80	63	22	230
Older people	26	37	27	13	103
TOTAL	91	117	90	35	333
THEORETICAL STAFFING TABLE					
Products	Rice	But	Peanut	Yam	TOTAL
Adults	62,85	80,81	62,16	24,17	230
Older people	28,15	36,19	27,84	10,83	103
TOTAL	91	117	90	35	333
CALCULATED CHI-SQUARE TABLE					
Products	Rice	But	Peanut	Yam	TOTAL
Adults	0,07	0,01	0,01	0,20	0,29
Older people	0,16	0,02	0,03	0,44	0,64
TOTAL	0,24	0,03	0,04	0,63	0,93
Degree of freedom		(l-1)*(c-1)			
Degree of freedom		3			
Significance threshold			5%		
Theoretical chi-square			7,82		

Source: Fieldwork, 2022-2023

The diagnosis in Table 4 indicates that the calculated Chi-square is smaller than the theoretical Chi-square. Numerically, 0.93<7.82, so the age of producers has no influence on food production. This population of food producers in the research framework is largely made up of indigenous Senufo.

I-4-Food crop farming mainly practised by indigenous people

The Senoufo people, originally from the region, are the landowners in the research area, even though they share their land with other peoples. The field survey revealed that they are the main labourers for food crops. Figure 5 highlights the quality of the workforce used for this type of farming.

Figure 5: Nationality of food crop producers

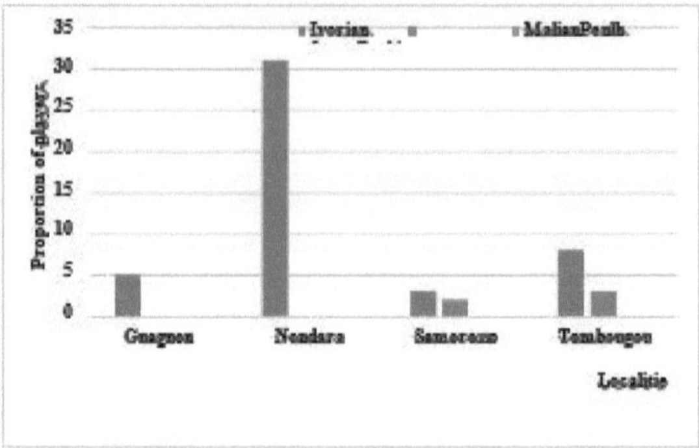

Source: 2022-2023 field survey

Figure 5 shows the preponderance of workers of Ivorian origin, and Senoufo in particular, in the 4 localities surveyed. However, the survey also showed that the localities of Samorosso and Tombougou have a workforce of Malian origin. The proportion of exogenous workers is as follows: 3.85% in Samorosso and 5.77% in Tombougou. It should be noted that Nondara and Gnagnon each have 100% Ivorian heads of household. These analyses show that the 4 localities in the study block are not the preferred destinations for migrants wishing to farm. In addition, a large proportion of the foreign population who settle there are involved in livestock rearing and trade. Figure 5 shows that 90.38% of the population in the 4 rural localities in the research area is indigenous, compared with 9.62% Malian. Senoufo farmers are also very keen on traditional marriage.

I-5 Households of food producers

I-5-1-Couple: farmers-customary marriages

Food-producing farmers are very sensitive to customary marriages. Figure 6 shows that only Nondara has single heads of household. Thus, 12.90% of heads of household in Nondara have no wife, compared with 87.10% of heads of household who have at least one wife. In the 4 localities in the research area, 81.82% of heads of household are married under customary law, while 18.18% are unmarried. Married heads of household are either monogamous or polygamous. The survey revealed that more than 50% of heads of households married under customary law have custody of at least two (2) wives, whereas less than 50% have a single spouse. The predominance of customary polygamous marriages to the detriment of monogamous marriages could be justified by work in the fields, as one respondent put it: 'Work in the fields requires a lot of people, hence the need for us to marry several wives'. Figure 6 shows the proportion of married and single heads of household. The children of these farmers go to school, while the income generated by their activity is insignificant.

Figure 6: Marital status of heads of farming households in localities

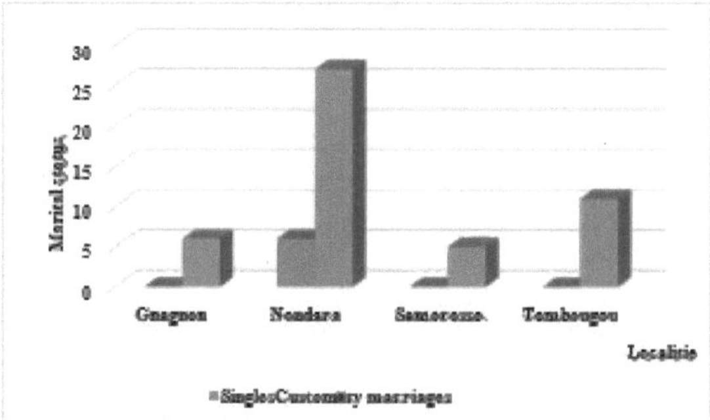

Source: Fieldwork, 2022-2023

I-5-2- Low school enrolment among the offspring of food crop producers.

Farmers send their children to school. Looking at Table 5, the school enrolment rate for children of farmers in the study area varies between 33.33% in Gnagnon and 51.85% in Tombougou, according to the data in Table 26. Samorosso and Tombougou have rates above 50%, and even higher than that of the study area, estimated at 43.50%. On the other hand, Gnagnon and Nondara have rates below the index for the research area (43.60%). The heads of farming households in the 4 localities surveyed enrolled a total of 143 children out of 328. On the other hand, across the whole study area, Gnagnon, Samorosso and Tombougou had average enrolment rates of 4.88%, 6.71% and 12.80% respectively. In Nondara, the enrolment rate for the 4 localities surveyed was 19.21%.

Table 5: Number of children of food producers attending school

Localities	Number of children in charge	Number of children in school	Percentage of children enrolled (%)
Gnagnon	48	16	33,33
Nondara	156	63	40,38
Samorosso	43	22	51,16
Tombougou	81	42	51,85
TOTAL	328	143	43,60

Source: 2022-2023 field survey

Figure 7 shows the different proportions of children attending school among farmers-producers of food crops according to the localities studied.

Figure 7: Number of children of food producers attending school

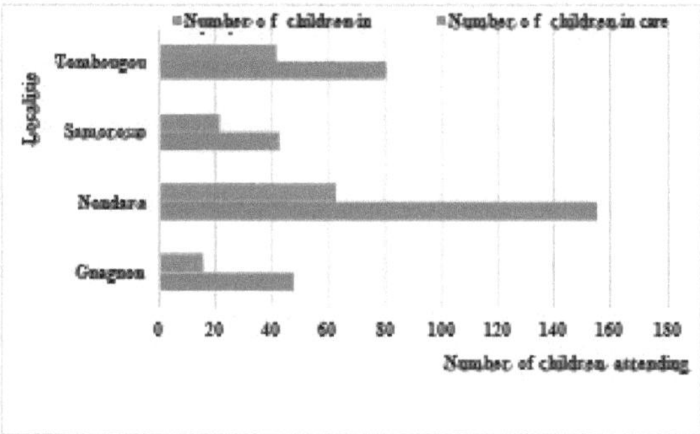

Source: Fieldwork, 2022-2023

An analysis of Figure 7 shows that food producers send few of their children to school. This finding could confirm the failure of national policies on compulsory enrolment for all children of school age, supported by free schooling in the study area. Also, the use of children as labourers (photo 2) for food production could justify this under-schooling.

Photo 2: Young teenager driving oxen for food production in Samorosso

Source: SOUMAHORO photo, 2022

The socio-demographic profiles of food crop growers vary. They provide a better understanding of food production in terms of externalities.

II- Characteristics of local food production in Boundiali

II-1-Conditions relating to the food producer's land

II-1-1-Mode of land acquisition dominated by land inheritance

Tenancy and sharecropping are very limited forms of access to land. In Nondara, 54.55% are land heirs and 47.27% of heads of household have access to land through gifts. In addition, 1.82% of producers rent land and 3.64% share production as a means of accessing agricultural land. In Samorosso, the proportion of land donated and sharecropped is very low, with 7.27% using the inheritance method. However, food producers in Gnagnon and Tombougou are exclusively land inheritors (Figure 8). In all, 92.73% of farmers inherited land. On the other hand, farmers renting land in the four localities pay an amount of between €9.15 and €22.87 to the landowners in proportion to the area cultivated during the year. This annual rental of agricultural land is due to the arid climate of the study block, which imposes a single production season for food crops. However, in order to anticipate and prevent any potential conflicts over the land, tenants of agricultural land are prohibited from planting perennial crops.

Figure 8: Number of farm households by type of access to land

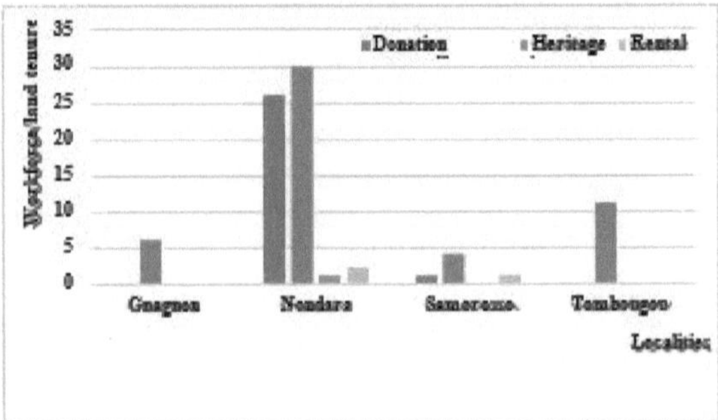

Source: 2021-2022 field survey

In addition to access to land, food production in the research area depends on the use of agricultural inputs.

II-1-2-Food production dependent on agricultural inputs

Food production requires a considerable amount of fertiliser, given the poor nutrient status of the soil. Table 6 shows the different types of input used in the study block.

Table 6: Operating account for a hectare of maize or rice under fertiliser in the study area

Designation	Expenses		
	Quantity	Unit price (€)	Amount (€)
NPK fertiliser (kg)	200	0,76	152,46
Urea fertiliser (kg)	50	0,91	45,74
Gramossone (litre)	1	7,62	7,62
TOTAL			205,82

Source: 2021-2022 field survey

Table 6 shows that three types of agricultural chemical are used in maize or rice production: NPK (Nitrogen, Phosphate, Potassium), Urea and Gramossone. Fertilising one hectare of maize or rice requires a total expenditure of €205.82, taking into account the figures in Table 2. A kilogram of NPK costs €0.76, or €38.16 per 50 kg bag, with 4 bags of NPK for one hectare. The cost of the 4 bags of NPK is €152.46, or 74.07% of total expenditure. Nitrogen-Phosphate-Potassium is a chemical that plants need for growth. Most of the soil's NPK requirements are met by organic inputs and by the soil, according to the Anader and Ivoire Cotons specialists interviewed. Urea, on the other hand, ensures the vegetative growth of the plant through the development of the stem and leaves. It accelerates the plant's growth rate. The cost of urea to maintain one hectare of maize or rice is one 50 kg bag, which accounts for 22.22% of total expenditure, or €45.74. Finally, Gramossone is a herbicide used to prepare the area to be sown. The agricultural technicians interviewed maintain that for one hectare of maize or rice to be cultivated, one litre of Gramossone is needed, which costs €7.62, or 3.71% of the financial expenses incurred. 51 respondents, or 92.73%, use fertiliser, compared with 7.27%, or 4 heads of farming households. Photos 3 and 4 show two agricultural fertiliser shops in Nondara.

Photo 3: Urea in a shop in Nondara

Photo 4: NPK in a shop in Nondara

Source: SOUMAHORO photo, 2022

Although fertilisers boost food production in the study area, labour is the lifeblood of the sector.

II-1-3-Farm labour mainly from outside the family unit

The results of our research show that 87.60% of respondents use outside labour, while 11.40% use family members. The use of foreign labour was associated with daily expenses of €2.29 per person. In addition, 45.1% of farmers surveyed used oxen, compared with 10.8% who used tractors. engines. An analysis of Table 7, which sets out the labour costs for developing a hectare of maize or rice, shows that hiring two oxen costs €30.49/hectare, while using a motorised machine costs €5.34/hectare. There are also more and more self-help farming groups, which are a form of assistance between food producers for €15.25/day. It is clear that, in terms of cost, self-help groups are more beneficial to farmers than isolated labour. On the other hand, mastery of cultivation techniques and methods would result in decent food production.

Table 7: Operating account for a hectare of maize or rice

Designation	Expenses		
	Quantity	Unit price (€)	Amount (€)
Hire a pair of oxen	1	30,49	30,49
Hire of motorised equipment	1	53,36	53,36
Daily labour cost	1	2,29	2,29
Daily cost of self-help group	10	15,25	15,25
TOTAL			101,39

Source: 2021-2022 field survey

II-1-4-Cultivation techniques dominated by manual farming

Cultivation practices in the research block can be broken down into three techniques: harnessed farming, manual farming and motorised farming (Figure 9).

Figure 9: Share of cultivation techniques used by rural producers

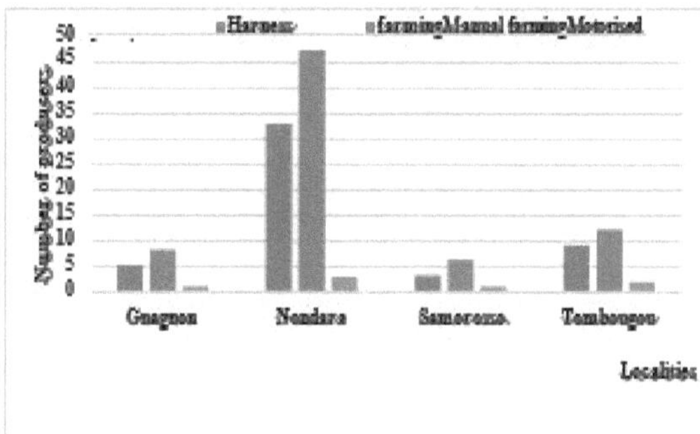

Source: Fieldwork, 2022-2023

The analysis in Figure 9 shows that harnessed farming and manual farming are the most popular. Thus, 38.46% of the farmers surveyed use harnessed farming, compared with 56.15% who use manual farming. Motorised farming was used by 5.39% of farmers. The graph shows that most food crop production in our area uses traditional tools. In other words, food crop production depends on rudimentary cultivation techniques. The archaic nature of the means of production could have a negative impact on productivity. Photos 5 and 6 show the use of animal-drawn cultivation. in Samorosso and manual farming in Nondara. However, the maize and rice fields are more budget-intensive than the others. All of these production factors require a budget for proper implementation.

Photo 5: Harness farming in Samorosso

Photo 6: Manual farming in Nondara

Source: SOUMAHORO photo, 2022

II-1-5-Low production costs for groundnuts

Maize (Zea mays) and groundnuts (Arachis hypogaea) are the main food crops grown in the research area. The financial investment made by the farmer to maintain these two crops is not negligible. An examination of Table 8 reveals a high production cost for maize, to the detriment of groundnuts. The investment cost of producing one hectare of maize is €307.21, compared with €28.97 for groundnuts, a difference of €278.24. Groundnuts do not require chemical fertilisers for their growth. The plant has the capacity to produce its own nutrients, particularly nitrogen. Farmers therefore combine the two crops in the same area. This advantage for groundnuts encourages 85.45% of farmers to grow them. Funding for food production in the research framework comes from a variety of sources.

Table 8: Farming account for one hectare of maize and groundnuts for one farmer

Crops	Type of workforce	Labour costs (€)	Nature of agricultural inputs	Farm input costs (€)	Land rental (€)	TOTAL (€)
Maize	1 pair of steers	30,49	NPK fertiliser (kg)	152,46		
	3 Manoeuvres	6,86	Urea fertiliser (kg)	45,74		
		30,49	Gramossone (litre)	7,62		
	2 group laps self-help		4 boxes herbicides	24,39		
TOTAL (€)		67,85		230,22	9,15	307,21
Peanut	1 group round self-help	15,25				
	2 Manoeuvres	4,57				
TOTAL (€)		19,82			9,15	28,97

Source: Fieldwork, 2022-2023

II-1-6 Source of funds for food crop production dominated by self-financing in Boundiali

Food production in Boundiali is mainly financed by the producer's purse. Analysis of Table 9 shows that 94.54% of respondents self-finance their food production. On the other hand, 5.45% of households surveyed take out loans from microfinance institutions: the Boundiali COOPEC. The amount of the loans varies between €152.46 and €457.39. It should be noted that the financial resources generated by the sale of cotton, cashew nuts and mangoes are used to finance food crops.

Table 9: Financing of small-scale food production in localities

Localities	Self-financing	Loan	Credit	TOTAL
Gnagnon	6	0	0	6
Nondara	33	0	0	33
Samorosso	5	0	0	5
Tombougou	8	3	0	11
TOTAL	52	3	0	55

Source: 2022-2023 field survey

II-1-7-Farmers producing food crops in Boundiali left to their own devices

The level of supervision of farm households in Boundiali is very low, according to Table 10. In fact, 39 heads of farming households are left to their own expertise, i.e. 70.91%, compared with 16 producers surveyed who receive advice and supervision from ANADER, i.e. 29.09%. The analysis in Table 10 shows that Nondara has the highest number of heads of farming households who are not supervised by ANADER (31, or 93.94%), compared with 79.49% of the farmers surveyed in the 4 localities. Lastly, Gnagnon is characterised by a very low level of supervised producers, with 4 heads of household, or 7.27%.

Table 10: Level of producer support

Localities	Supervised households	Unregulated households	TOTAL
Gnagnon	2	4	6
Nondara	2	31	33
Samorosso	5	0	5
Tombougou	7	4	11
TOTAL	16	39	55

Source: 2022-2023 field survey

As well as food crops, pastoral activities play a key role for Senufo farmers in this northern part of Côte d'Ivoire.

II-2-Agropastoral: an ancillary activity to subsistence farming

The region is traditionally suited to all types of livestock farming: cattle, sheep, goats, pigs and poultry. Traditional livestock farming is widespread. It is characterised by an intensive system, increasingly sedentary or semi-transhumant. This type of farming poses problems of cohabitation between Fulani herders and Senufo or Malinke farmers.

II-2-1-Conditions for the development of livestock farming

The vegetation and climate in the research area are suitable for animal protein production. The plant cover is essentially savannah. The climate is characterised by relatively low humidity. The proximity of the study area to the Sahelian countries of West Africa means that it has a livestock farming culture. This geographical advantage in favour of pastoral activity is reinforced by the existence of dams or water reservoirs, as shown in Table 11.

Table 1: Number of water reservoirs in the Boundiali area by purpose

Zone	Vocations					Total
	AEP	Cane	Breeding	Multiple	Rice	
Boundiali	-	-	50	1	1	52

AEP : *Drinking Water Supply

Source: SILUE Pébanagnanan David, 2012

Analysis of Table 11 reveals the presence of 52 dams in the Boundiali area out of 72 in the Bagoué region and 275 reservoirs in the Savanes district. The Boundiali area therefore accounts for 72.22% of regional dams and 18.90% of the number of reservoirs in the district. Only 12 dams have been developed in Boundiali out of a total of 52, representing an implementation rate of 23.07%, according to the results of surveys of the authorities in charge of managing water reservoirs. Examination of Table 31 shows that 50 of the 52 dams, or

72.46%, are used as pasture in Boundiali. However, one dam covers our geographical area: Tombougou-Samorosso in Boundiali, i.e. 1.92% of the total number of dams in the department, as shown in photo 7.

Photo 7: Tombougou-Samorosso dam, 3 km from Boundiali

Source: SOUMAHORO photo, 2023

The potential for livestock development in the research area is undeniable. They led to the implementation of various hydraulic pastoral projects in the Boundiali area between 1976 and 1990, as shown in Table 12.

Table 2: Pastoral hydraulics project in the Boundiali area

Projects	Period of performance	Zone
Operation Zébu	1976-1982	Boundiali
Operation Taurin	1976-1982	Boundiali
Pastoral developments	1982-1990	Boundiali

Source: KOUADIO.D. B, 2004

Table 12 shows that from 1976 to 1982, zebu and bull operations were carried out in Boundiali, as well as pastoral developments. This state of affairs suggests that the Boundiali environment is conducive to the development of pastoral activity. However, this economic activity in the primary sector is struggling to get off the ground, given the difficulties encountered by those involved. Conditions for food crop production remain difficult, given the bottlenecks that mark out the activity. These bottlenecks could act as tentacles that put the brakes on production.

III- Presentation of food crops

III-1-Areas sown by farmers surveyed, with Nondara predominating

Nondara is the preferred area for growing food crops in the commune of Boundiali.

Table 13 attests to this fact.

Table 13: Area sown to food crops in hectares in Boundiali

Localities	Rice	Maize	Mil	Sorghum	Beans	Peanut	Yam	TOTAL
Gnagnon	15	19	3	3	1	11	5	57
Nondara	59	116	6	0	0	144	13	338
Samorosso	11	10	0	0	0	4	2	27
Tombougou	19	32	0	0	1	14	1	67
TOTAL	104	177	9	3	2	173	21	489

Source: Field surveys 2021-2022

Maize, groundnuts and rice occupy large areas, with 177 ha (36.20% of the total area sown), 173 ha (35.39%) and 104 ha (21.27%) respectively (Table 13). In addition, the area sown to yam, at 21 ha (4.29%), is higher than that sown to millet (9 ha, or 1.84%), sorghum (3 ha, or 0.61%) and beans (2 ha, or 0.41%). Samorosso has the smallest cultivated area, accounting for just over 20ème (27 ha or 5.52%) of the sown crop cover. The small area observed at Samorosso may be due to the numerous housing developments undertaken by the community, which de facto reduce the area available for cultivation. Also, access to land in this locality is very rigid. Thus, renting and sharecropping are very limited forms of access to land, especially in Samorosso. 54.55% of farmers in Nondara are land heirs, compared with 47.27% who access land by gift. In addition, 1.82% of farmers rent land and 3.64% share production as a means of using arable land. On the other hand, donation and sharecropping account for a very small proportion (7.27%), in favour of access by inheritance in Samorosso. In addition to these bottlenecks, there is the exodus of skilled workers to the town of Boundiali to work in the secondary and tertiary sectors, and the urbanisation of Boundiali to the detriment of Samorosso, located barely three km from the town, limiting food production through a drastic reduction in arable land. On the other hand, these markers marking out access to agricultural land could anticipate and prevent any desire for land disputes. In addition, the area sown could explain the quantity of food crops harvested.

III-2-Local food production dominated by maize and groundnuts

Maize (Zea mays), groundnuts (Arachis hypogaea), rainfed and lowland rice (Oryza sativa), millet (Pennisetum glaucum), sorghum (Sorghum bicolor), beans (Phaseolus vulgaris) and yams (Dioscorea spp) are the main food crops grown in the study block (Table 14). Specifically, maize and groundnuts dominate vegetable food products, with annual production of 185,604 kg (37.85% of total annual production) and 18,3300 kg (37.38%) respectively, as shown in Table 3. Nondara is the main locality that could supply the town of Boundiali, given its agricultural statistics: 148400 kg of groundnuts (41.44%), 129000 kg of maize (36.02%) and 76500 kg of rice (21.36%). Photo 8 shows how groundnuts are grown in Nondara. However, the first two crops are the food crops grown in Boundiali, to which rice should also be added. The predominance of maize could be justified by its compatibility with the climatic regime of the area. In addition, the subjection of the other crops can be explained by the the diet of the Senoufo people, which is essentially based on maize. Finally, the accessibility A plausible reason for this dominance would be the fact that consumers have a lower price for this product.

Table 14: Annual food crop production in kilograms

Localities	Rice	Maize	Mil	Sorghum	Beans	Peanut	Yam	TOTAL
Gnagnon	11000	16100	1800	750	100	11200	1000	41950
Nondara	76500	129000	0	0	0	148400	4200	358100
Samorosso	7500	9500	0	0	0	4200	2000	23200
Tombougou	14500	31004	0	0	0	19500	2000	67004
TOTAL	109500	185604	1800	750	100	183300	9200	490254

Source: 2021-2022 field survey

Photo 8: Groundnut field at Gnagnon, 3 km from Boundiali

Source: SOUMAHORO photo, 2022

The areas sown and the volume of production account for good or poor farm yields.

III-3-Better yields for food crops in Tombougou

Groundnuts, maize, yams and rice have high yields compared with the other crops, with gross yields of around 4491.58 kg/ha, 3878.29 kg/ha, 3523.07 kg/ha and 3474.90 kg/ha respectively (Table 15). A comparative analysis of these yields shows that groundnuts have the highest yield: 4491.58 kg/ha, or 27.53% of total yields. Local food production supplies urban markets. The high yield of food crops in Tombougou could be explained by the quality of the nutrient-rich soil and the farmers' compliance with farming methods and techniques. Livestock farming is added to these food production activities.

Table 15: Yields in kg/ha of food crops

Localities	Rice	Maize	Mil	Sorghum	Beans	Peanut	Yam	TOTAL
Gnagnon	733,33	847,36	600	250	100	1018,18	200	3748,87
Nondara	1296,61	1112,06	0	0	0	1030,55	323,07	3762,29
Samorosso	681,81	950	0	0	0	1050	1000	3681,81
Tombougou	763,15	968,87	0	0	0	1392,85	2000	5124,87
TOTAL	3474,90	3878,29	600	250	100	4491,58	3523,07	16317,84

Source: 2021-2022 field survey

III-4-Animal production to the advantage of chickens in Boundiali

Livestock production is dominated by chickens of African origin known as "bicycle chickens". Cattle (Bos taurus) are in second place, followed by sheep (Ovis aries) in third. Goats (Caprinae) are in fourth place. The breeding of guinea fowl (Numididae) and pigs (Sus scrofa domesticus) is not developed according to the data in table 16.

Table 16: Heads of livestock per year by type of farming

Localities	Chickens	Cabris	Sheep	Steers	Pigs	Guinea fowl	TOTAL
Gnagnon	23	12	0	9	12	0	56
Nondara	264	60	169	181	0	30	704
Samorosso	17	0	25	36	0	0	78
Tombougou	38	6	32	31	0	0	107
TOTAL	342	78	226	257	12	30	945

Source: 2022-2023 field survey

Table 16 shows that 342 chickens (36.19%), 257 cattle (27.19%) and 226 sheep (23.69%) are reared. Goats and guinea fowl are reared in insignificant numbers: 78 head, or 8.25% of goats, and 30 head of guinea fowl, or 3.17%. Finally, pig farming is almost non-existent, with only 12 head, or 1.27% (photo 9). However, table 16 shows that Nondara is the area with the strongest tradition of livestock farming, with 704 head of livestock (74.50%), including 37.5% chicken, 8.52% goats, 24.01% sheep, 25.71% beef and 4.26% guinea fowl. All in all, livestock production in the rural areas of Boundiali is very inadequate. Chicken farming is at the top of the peleton. These various local food products are transported to the Boundiali market to supply consumers.

Photo 9: Three pigs in the village of Gnagnon, 2 km from Boundiali.

Source: SOUMAHORO photo, 2023

IV- Supplying the urban commercial centre of Boundiali with local food products

IV-1-Storage facilities for food products, dominated by warehouses in Boundiali

Granaries and warehouses are the main infrastructures for storing food products. Analysis of Figure 10 shows that 80.52% of respondents keep their produce in shops built with modern materials, while 19.48% store it in granaries. However, 70.13% of producers prefer to shop their produce in village warehouses, compared with 10.39% who take their produce to the town for storage. This shows the resilience of the granary in the food production chain in the face of the accelerated modernisation of rural housing.

Figure 10: Places where food products are stored

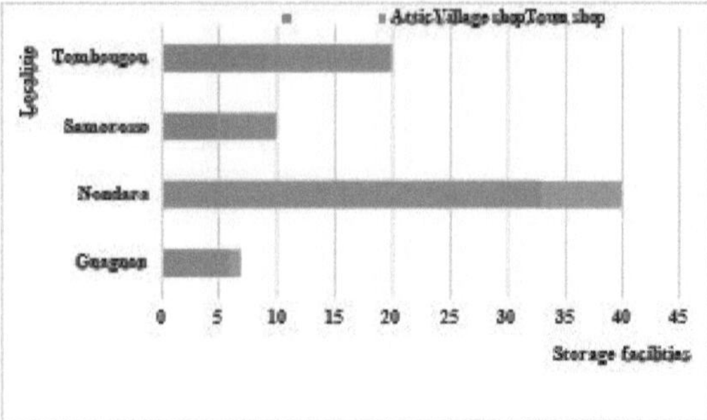

Source: Field work, 2021-2022

Photo 10 shows the existence of this traditional infrastructure for storing food products in Gnagnon. Goods are transported from these warehouses to the Boundiali market.

Photo 10: Granary for storing maize at Gnagnon

Source: SOUMAHORO photo, 2022

IV-2-Transport means for the sale of local food products

Two- or three-wheeled motorbikes, tarpaulins weighing 2 to 9 tonnes and mansa with 18 to 22 seats are used to transport 94.68% of rural food crops (Figure 11). Thus, 23.89% of rural producers transport their produce in tarpaulins, compared with 22.12% in public transport vehicles, 48.67% in two- or three-wheeled vehicles, 2.65% on their heads, 0.9% in bicycles and 1.77% in pack animals. The high use of two- or three-wheel motorbikes is explained by the low volume of local food production to be transported from the rural production basins to the town of Boundiali.

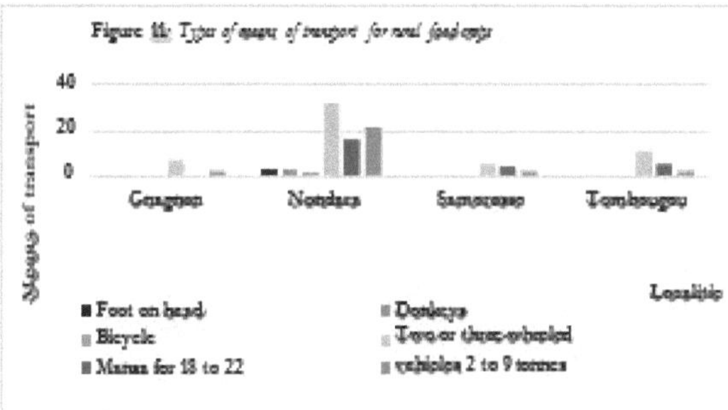

Source: Field work 2022-2023

Figure 11 shows that three-wheelers are the main means of transporting food crops, as Photo 11 also shows.

Photo 11: Three-wheeled machine loaded with food crops in Boundiali

Source: SOUMAHORO photo, 2022

IV-3-Relationship between local food production and the Boundiali market.

IV-3-1-Relationship between food crop production and the Boundiali market.

Nondara markets 46900 kg of maize, 106600 kg of groundnuts and 23900 kg of rice, and remains the food granary of the study area (Figure 12). However, this production (490254 kg per year) remains insufficient for an urban population of around 65191, i.e. 7.52 kg/head/year. In addition to the plant foodstuffs found in the transaction area, consumers are also interested in animal foodstuffs.

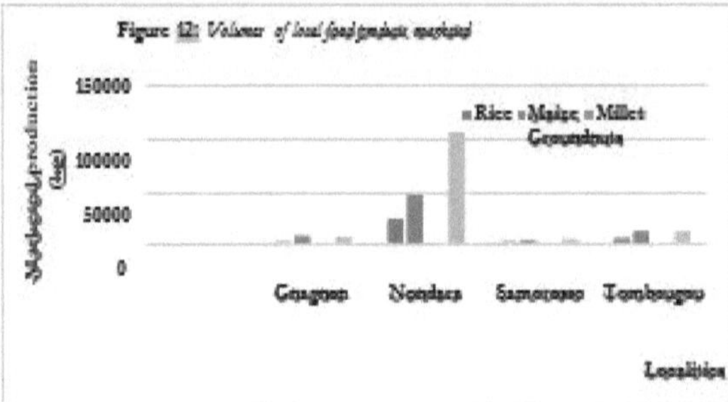

Source: Fieldwork, 2022-2023

IV-3-2-Relationship between food animal production and the Boundiali market.

Oxen, chickens, sheep, goats, guinea fowl and pigs are the livestock products sold by producers at the Boundiali market. Figure 13 shows that chickens, oxen and sheep are the most popular products on the Boundiali market. Chickens account for 40.93% of animals sold, compared with 24.60% for oxen and 21.98% for sheep. Goats, guinea fowl and pigs are sold to a lesser extent, accounting for 8.67%, 3.02% and 0.81% respectively. Nondara is the biggest supplier of animal products, with 79.44% of animals sold, while Tombougou (10.08%), Samorosso (6.45%) and Gnagnon (4.03%) are struggling to supply large numbers of animals to meet the protein needs of an estimated 6,5191 urban consumers. Ultimately, the four localities provide the Boundiali market with a total annual animal production of 496 head, or 0.008 head/head/year. In addition to the low animal protein production, the market prices vary.

Figure 13: Heads of local animals sold at the Boundiali market per year

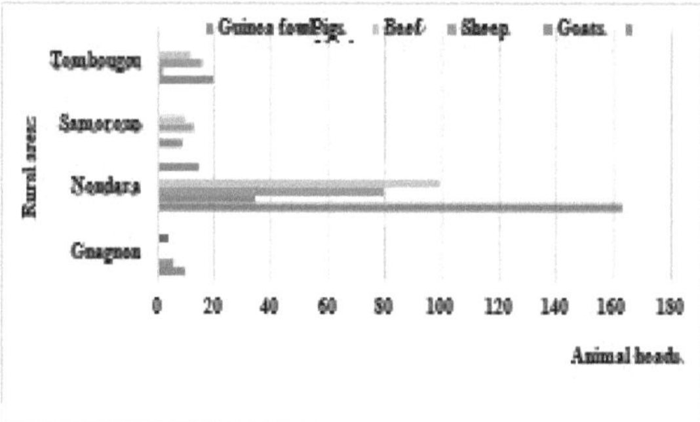

Source: Fieldwork, 2022-2023

IV-4- Prices of food products sold on the Boundiali market

IV-4-1- Prices of marketed food products with a high market value of Peanuts

Groundnuts cost €0.99 per kg, compared with €30.49 per 100 kg bag of unmashed rice, maize €26.68 per 100 kg bag and millet €0.30 per kg (Table 17). Looking at Table 17, groundnuts have a high market value at €0.99/kg. To make the transaction possible, local units of measurement are used, which are the bucket, the heap and the box. A box of maize costs €0.27, while a box of millet costs €0.30. Photo 12 shows the types of food products available at the Boundiali market. The high market value of groundnuts could support producers' choice to grow this crop.

Table 17: Average selling price of food crops at producer level

Food crops	Quantity	Unit price (€)	Amount (€)
Unmashed rice (kg)	100	0,30	30,49
Maize (kg)	100	0,27	26,68
Millet (kg)	100	0,30	30,49
Groundnuts (seed in kg)	100	0,99	99,10

Source: Field work 2022-2023

Photo 12: Cereals sold at the Boundiali market

Source: SOUMAHORO photo, 2022-2023

Looking at photo 12, crushed maize, white rice, beans, millet and sorghum are the most popular food crops at the Boundiali market. Maize and groundnut production generate income for farmers. In fact, maize is more expensive to grow than groundnuts (Table 18). Looking at Table 18, food producers need to raise €307.21 to grow one hectare of maize, compared with €28.97 for groundnuts. Groundnuts, on the other hand, are more profitable for the farmer because, for a total production cost of around €28.97 per hectare, the farmer makes an annual profit of €767.98, more than double the profit made from the sale of maize. As well as plant-based food products, the selling prices of animal proteins depend on the type of animal.

Table 18: Annual operating statement for a hectare of maize and groundnut fields from a farmer in Nondara

Crops	Maize	Total (€)	Peanut	Total (€)
Rental cost of the earth (€)	9,16	9,16	9,16	9,16
Nature of the hand work	1 pair of oxen 3 manoeuvres 2 group laps self-help		1 group round self-help 2 manoeuvres	
Hand cost of work (€)	67,92	67,92	19,84	19,84
Agricultural inputs	NPK fertiliser (kg) Urea fertiliser (kg) Gramossone (litre) 4 boxes herbicides			
Input costs agricultural (€)	230,48	230,48		
Total expenditure (€)		307,56		29
Production volume sold (kg)		2500		803,30
Selling price of revenue (€)		667,79		796,99
Earnings (€)		360,22		767,98

Source: Field work, 2021-2022

IV-4-2 Prices of marketed food and animal products with a market value high beef

Livestock prices in the research area differ from one animal to another. Poultry products: guinea fowl and chicken have very low average selling prices, with €5.34 for guinea fowl and €4.57 for chicken, a difference of 7.69% in favour of the former (table 19). The prices of sheep, pigs and goats have risen in relation to these prices. Table 19 shows that sheep are sold at €53.36, pigs at €38.16 and goats at €30.49, giving an average unit price of €40.66. Beef costs more than any of the above. Looking at table 19, the average unit price of beef is €251.56, a difference of 72.17% and 8 times the price of the aforementioned animals. More than 80% of the producers interviewed maintain that these prices are not always respected by the traders who buy the animals. Photos 13 and 14 show the transaction of guinea fowl and beef (€3.81/kg) in the Boundiali market. Trade in local food products is a source of financial resources for the Boundiali town council.

Table 19: Average selling price of food animal products at producer level

Designation	Quantity (s)	Average price (€)
Chicken (s)	1	4,57
Guinea fowl (s)	1	5,35
Cabri (s)	1	30,49
Sheep (s)	1	53,36
Beef (s)	1	251,56
Pig (s)	1	38,16

Source: Field work 2022-2023

Photo 13: Guinea fowl at the Boundiali market **Photo14:** Beef at the Boundiali market

Source: SOUMAHORO photo, 2022

IV-5-Contribution of local food crops to the management of Boundiali town council

The revenue collected by Boundiali town council from 2017 to 2021 varies between €2,869.34 and €25,461.20 (Table 20). Looking at Table 20, the share of taxes in 2017 was €6459.61, or 27.59% of revenue, compared with €3819.12 in 2021, or 15.04% of revenue. Food traders pay €0.30 per day. The daily amount of taxes is €274.43 in 2021 for 900 places out of 1,002 places available in the market. This shows the importance of the food trade in the Boundiali town council's tax base.

Table 20: Revenue growth at the Boundiali market from 2017 to 2021

Years	Taxes (€)	Royalties (€)	Revenue (€)
2017	6459,61	16950,98	23410,58
2018	3808,05	20261,32	24069,07
2019	3532,17	20126,01	23658,18
2020	4930,28	19179,03	24109,32
2021	3819,12	21580,48	25399,60
Average over 5 years	4509,84	19619,57	24129,41

Source: Fieldwork, 2022-2023

Figure 14 shows that Nondara remains the leading supplier to the town of Boundiali.

Figure 14: Flows of rural food products to the town of Boundiali

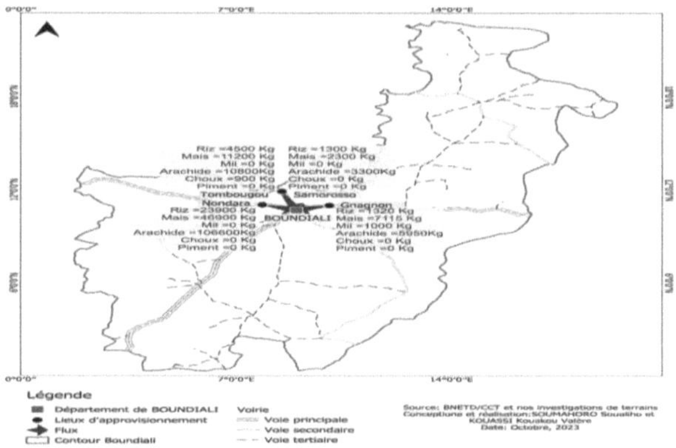

Source: Fieldwork, 2022-2023

The marketing system comprises two levels: the short circuit and the long circuit. An analysis of Table 21 shows that 64 women are involved in marketing food crops (photo 15).

Table 21: Number of women producers selling food crops in the study block

Localities	Number of women
Gnagnon	08
Nondara	34
Samorosso	07
Tombougou	15
Total	64

Source: Field work, 2021-2022

Photo 15: Women farmers delivering food to traders in the Boundiali market.

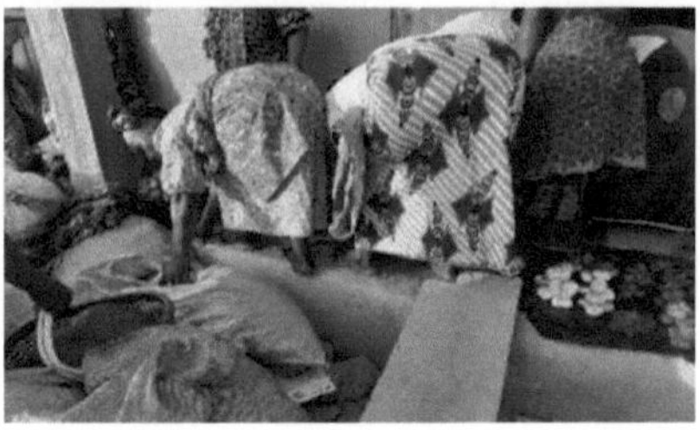

Source: SOUMAHORO photo, 2022

Figure 15 shows that 37.21% of women deliver their goods directly to consumers, compared with 62.79% who sell them to traders.

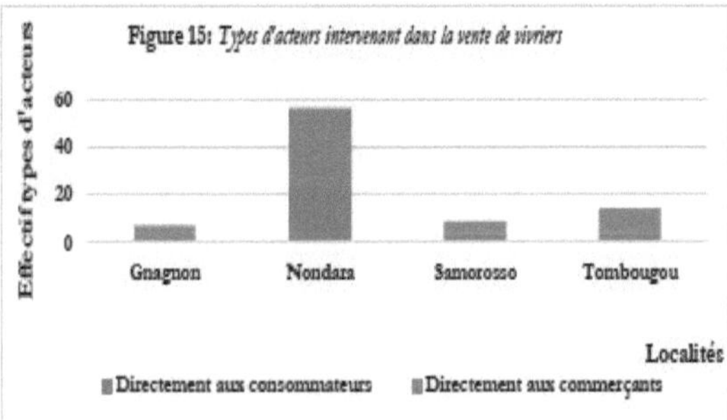

Source: Field work, 2021-2022

The long circuit puts the producer in contact with wholesalers, retailers and intermediaries. Figure 16 shows that 46.73% of producers sell their crops to wholesalers, compared with 48.60% to retailers and 4.67% to intermediaries.

The supply systems in the town of Boundiali are dominated by long indirect channels (Figure 16). However, it should be noted that farmers' wives offer food crops to consumers, especially on Saturdays, the highlight of the Boundiali market.

Figure 16: *Acteurs de la chaine de commercialisation des produits vivriers locaux*

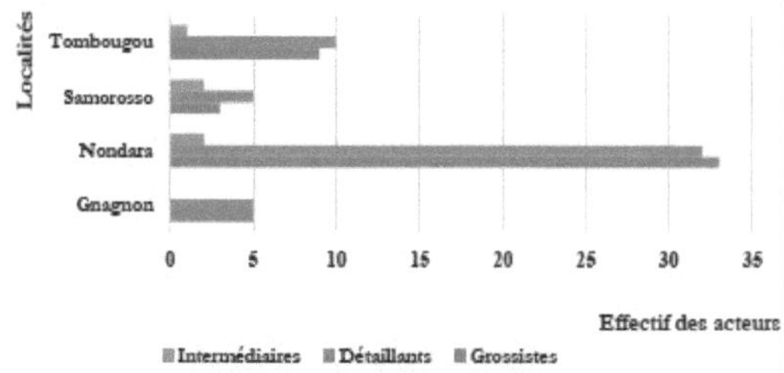

Source: Field work, 2021-2022

The Boundiali market generates additional resources for the town council to implement its programme. Food production also helps to improve the living conditions of producers in the research area.

IV-6-Social and economic impact of food production on farmers' well-being

IV-6-1-Advantage of local food production in producers' self-sufficiency

Table 22 shows that beans, yams and sorghum are consumed by farmers themselves, while groundnuts are grown mainly for sale, with only 30.91% for consumption. In addition, 63.62% of maize and 71.67% of rice are consumed by farmers (photo 16). An analysis of Table 2 shows that the 4 localities have an annual production of around 26,4069 kg, or 4,801.25 kg/farm household/year for consumption. By In addition, these four localities produce 226185 kg for sale in Boundiali, i.e. 3.47 kg/citadin/year for marketing to 65191 people. Specifically, 1,8089 kg of maize are produced for consumption in the village, i.e. 2,147.07 kg/farm household/year, whereas the ratio of maize produced in the town is 1.04 kg/head/year. The preponderance of groundnuts in the marketing chain of local food products could be explained by the monetary aspect for the schooling of their offspring and the physiological maintenance of their families.

Table 22: Food crop production in kg for consumption

Localities	Rice	Maize	Mil	Sorghum	Beans	Peanut	Yam	TOTAL
Gnagnon	9680	8985	800	750	100	5250	1000	26565
Nondara	52600	82100	0	0	0	41800	4200	180700
Samorosso	6200	7200	0	0	0	900	2000	16300
Tombougou	10000	19804	0	0	0	8700	2000	40504
TOTAL	78480	118089	800	750	100	56650	9200	264069

Source: Fieldwork, 2022-2023

Photo 16: Paddy rice stored in a warehouse in Gnagnon for self-consumption

Source: SOUMAHORO photo, 2022

In addition to food crops, producers also engage in pastoral activities. Table 23 shows that 30.33% of cattle, 31.24% of chickens, 26.29% of sheep, 6.97% of goats, 3.37% of guinea fowl and 1.80% of pigs are consumed in the study block. This analysis shows that cattle and poultry farming are the most developed. The dominance of cattle could be justified by the use of oxen for harness farming (photo 17) and for the organisation of rituals linked to Senufo customs.

Table 23: Heads of animals for consumption at producer level per year

Localities	Chickens	Cabris	Sheep	Steers	Pigs	Guinea fowl	TOTAL
Gnagnon	13	2	0	9	8	0	32
Nondara	100	25	89	81	0	15	310
Samorosso	8	0	12	26	0	0	46
Tombougou	18	4	16	19	0	0	57
TOTAL	139	31	117	135	8	15	445

Source: Fieldwork, 2022-2023

The low level of pork production (1.27% of total production and 1.80% of total consumption) could be explained by the Islamisation of populations where the consumption of this animal is forbidden by the aforementioned religion.

Photo 17: Hitch farming in Tombougou

Source: Fieldwork, 2022-2023

Food crops are mainly produced for self-consumption. However, the transaction of these products offers opportunities for practitioners to provide their households with capital goods.

IV-6-2-Impact of marketing local food products on the living conditions of Boundiali producers

Food producers partly improve their environment through the production and sale of food products. Figure 17 shows that 34.21% of income from the sale of food crops supports schooling for farmers' children. The foodstuffs sold also provide 19.74% of farmers with manufactured goods. Figure 17 shows that 10.09% of brave farmers use the funds generated to rehabilitate village pumps, compared with 27.19% for renting and buying houses in town and 8.77% for tontines. Photo 18 shows the significant impact of food transactions in providing households in Gnagnon with drinking water.

Figure 17: Living conditions of food producers in Boundiali

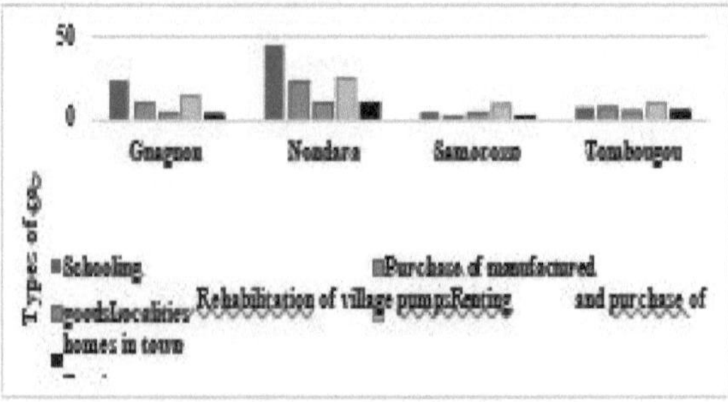

Source: Field work, 2021-2022

Analysis of Figure 17 shows that only 9.65% of producers in Samorosso benefit from the marketing of food crops. This could be explained, firstly, by the low volume of food production, secondly, by the strong presence of cotton and cashew nut crops and, thirdly, by the lack of a national food production policy, especially in terms of offering field prices to producers. Over and above the benefits of producing and distributing local foodstuffs in terms of positive changes to the living environment, this activity helps to maintain socio-cultural values.

Photo18: Functional human-powered pump at Gnagnon

Source: SOUMAHORO photo, 2022

IV-6-2-Retombées de la commercialisation des produits vivriers locaux dans la pérennisation des valeurs socioculturelles

The Senufo are a conservative people. Customs and traditions play a central role in their society. An analysis of figure 18 shows that funerals, sacred wood and weddings remain the main features of this people's civilisation. Thus, 34.68% of food producers devote a substantial proportion of their income to organising weddings. On the other hand, 33.53% of these producers finance their stay in the sacred grove through the sale of food crops, compared with 31.79% for funerals. To sum up, Figure 18 shows that socio-cultural values are the essential features of Senoufo civilisation in proportion to the proportions of these values.

Figure 18: Socio-cultural values in Boundiali

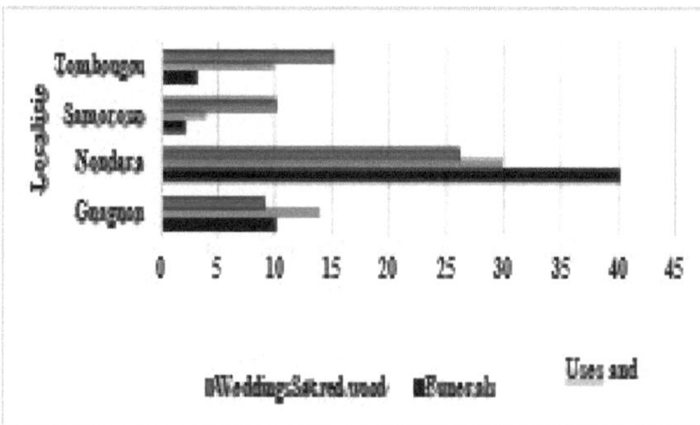

Source: Field work, 2021-2022

Photo 19 shows young Senufo being trained in the sacred forest at Nondara.

Photo 19: Two young Senufo undergoing training in the sacred grove at Nondarai.

Source: SOUMAHORO photo, 2022

Local food production supplies the Boundiali trade hub. Supplying this market generates income for the town council through the tax collection system. However, this meeting place for sellers, buyers and consumers may not be the best place to be.

V- Presentation of the Boundiali market

V-1 Boundiali market, the largest commercial centre in the Bagoué region

Located between latitudes 9°.52 and 9°.44 north and longitudes 6°.75 and 6°.47 west, the Boundiali market was built in 1985. It is rectangular in shape, 210 m long and 185 m wide (Figure 19). In fact, this commercial building covers an area of 38,850 m^2 with 10,002 seats, i.e. 3.90 m^2 per seat. This space is home to players from a variety of backgrounds.

Figure 19: Location of the Boundiali market

Source: Field work, 2021-2022

V-2-Social and demographic profile of food sellers

V-2-1- Women's food trade

More women than men are involved in the food trade. Figure 20 shows that 66.7% of respondents were women, compared with 33.30% of male traders in the Boundiali market.

Figure 20: Distribution of food traders by gender in the two markets

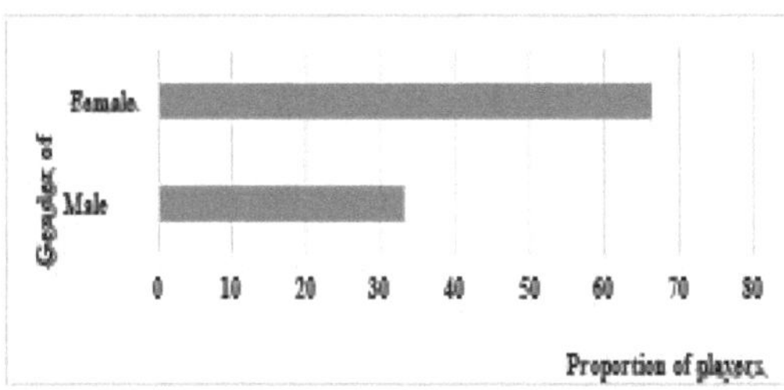

Source: Fieldwork, 2022-2023

What's more, the market on the research block is bustling with traders from a wide range of backgrounds.

V-2-2-Dominance of traders of Ivorian nationality in the sale of food crops

The market population in the study area is composite and cosmopolitan. However, 92.00% of traders are nationals, compared with 4.00% Malians and 4.00% Mauritanians (Figure 21). This area of commercial exchange is an excellent place for cultural mixing.

Figure 21: Breakdown of food traders by nationality

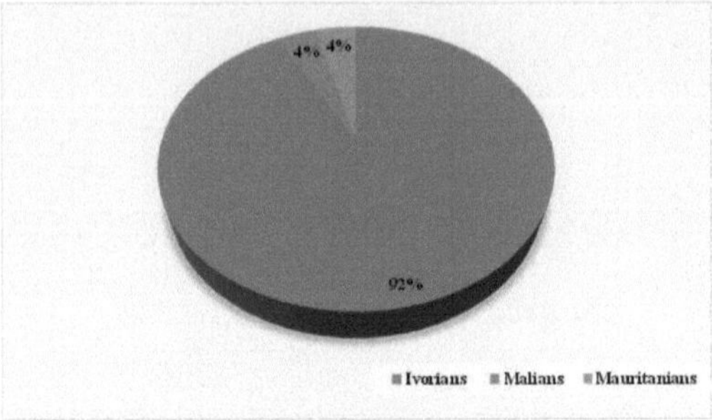

Source: Fieldwork, 2022-2023

Does this cultural diversity not influence the level of education?

V-2-3-Food trade in the hands of the uneducated

Those with no schooling, primary school and secondary school are the educational levels of the traders in this area. market. Figure 22 shows that illiterate people account for 75% of food traders (9 out of a total of 12), compared with 16.67% (2) who have completed primary school and 8.33% (1) who have completed secondary school. The socio-demographic characteristics of traders could have a positive or negative impact on trading activity.

Figure 22: Distribution of food traders by level of education

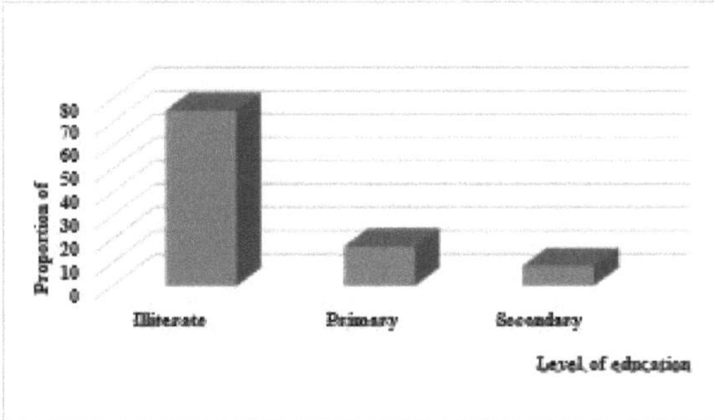

Source: Fieldwork, 2022-2023

V-3-Spatial distribution of food crops on the market

V-3-1-Disparate spatial distribution of food products in the Boundiali market

The central market in Boundiali offers users a varied range of food products. These products are concentrated in well-known, albeit disparate, sectors. The analysis in Figure 23 shows that food sellers display their goods on the periphery of the market. This organisation close to the roads is in response to the need to make the products accessible and visible to consumers. The results of the fieldwork revealed three levels of product exposure: agricultural foodstuffs, fish and animal products, and manufactured goods. Figure 23 shows that manufactured products are stored in warehouses. The display of foodstuffs next to alleyways could lead to unfair competition with regard to the payment of taxes. Many vendors refuse to pay taxes because of this situation, which drastically affects the council's revenue.

Figure 23: Spatial layout of foodstuffs in the Boundiali market

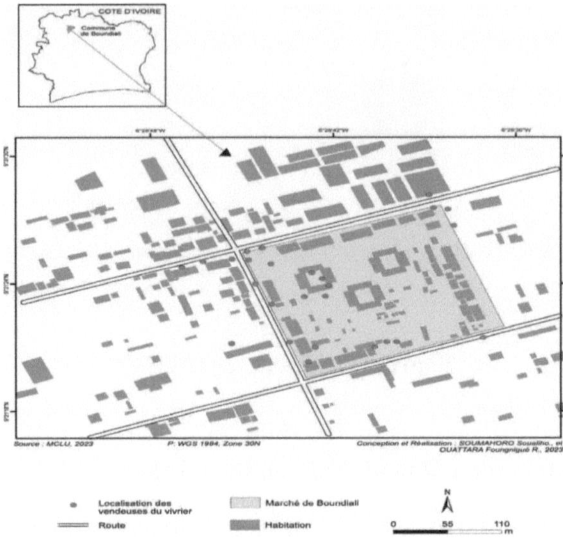

Source: Field work, 2021-2022

Saturday is the highlight of the Boundiali market. Sellers and buyers from all over the world gather there. As a result, the natural sales area becomes extremely narrow, forcing some traders to use the pavement to display their goods (photos 20 and 21).

Photo 20: Fish on a shelf

Photo 21: Vegetable sector at the Boundiali market

Source: Photo SOUMAHORO 2021-2022

The various shelves stocked by traders at the Boundiali market offer consumers a mosaic of foodstuffs, particularly fish products (photo 13). This The diversity of food products follows a structured organisation in terms of spatial layout. The deterioration of the market environment in the research block requires a detailed study for a thorough understanding of the phenomenon.

V-3-2-Specific approach to the unhealthy food sector in the Boundiali market

The environmental profile of the market in the study area was based on traders' perceptions. Table 24 shows traders' opinions on the hygiene of the food and manufactured goods sales area.

Table 24: Traders' perceptions of the state of the food sector environment in the study market

Boundiali market environment	Workforce	Frequency (%)
Own	0	0
Moderately clean	3	25,00
Very dirty	9	75,00
TOTAL	12	100

Source: Field work 2021-2022

Looking at Table 24, ¾ of the traders interviewed in the research block market maintain that the markets are very dirty and state that they are very affected by household waste, sewage, rainwater as well as bugs with their share of nauseating smells. On the other hand, more than ¼ of traders admit that the market is not too dirty. However, all the traders surveyed clearly stated that the market environment was not at all pleasant. In addition, 100% of the traders surveyed agreed that the areas dedicated to the sale of food products in the market in the study block were much worse than those used for the sale of manufactured goods. Examination of figure 24 reveals the unhealthy state of this public building. It can be seen that the area around the market is dirty. Photo 22 reveals the existence of stagnant waste water in clogged, sandy gutters and unauthorised rubbish dumps of all kinds. On closer examination, more than 75% of the vendors and buyers questioned said that the market's environmental conditions had deteriorated, compared with just 25% who said that it was relatively clean.

Figure 24: View of rubbish dumps in the Boundiali market

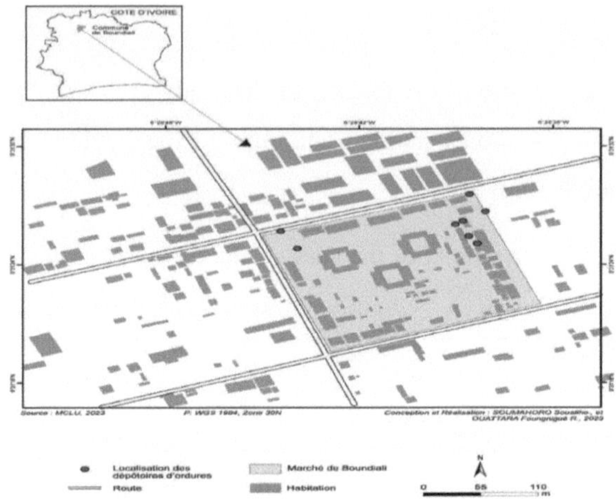

Source: Field work 2021-2022

This waste, which is creating an environmental disaster, is essentially made up of residues of marketed foodstuffs, although field work has shown that manufactured products make a significant contribution to the degradation of the environment (photos 22 and 23). The conclusion on the hygienic state of the Boundiali market environment is without paradox. It is an environmental hecatomb.

Photo 22: Blocked gutter at Boundiali market

Photo 23: Dump at Boundiali market Boundiali

Source: SOUMAHORO photo, 2023

DISCUSSION

Land rights are essentially customary, with access to agricultural land predominantly through inheritance. The order of succession is based on the patrilineal lineage. The kahafolo (village chief) or tarafolo (land chief) allocates plots of land to be farmed to family members according to need. Families then exercise an inalienable right of use. This land tenure practice de facto prohibits all forms of transaction. It appears that the nature of the method of access to land could constitute an obstacle to abundant food production. This reality observed in the research environment is supported by the work of DUFUMIER. M (1999 : 547-560). According to this author, undivided ownership of land is often seen as an obstacle to agricultural intensification. Access to land is therefore the most important barrier to entry into plantain production in the western region of Cameroon (Guillaume Hensel FONGANG FOUEPE et al., 2019: 5). Also, food production in the study area is held mainly by producers who cannot read, write or calculate. The low level of education of the players may act as a brake on mass production, according to the lines of research developed by Mathieu Juliot MPABE BOJONGO and Fondo Sikod (2017: 4). They justify that an increase in the level of education favours an increase in agricultural income. In their view, education can help to improve agricultural productivity through the easy appropriation of new agricultural techniques. Compared to the level of education dominated by the uneducated, older people are responsible for food production in the research block. The numerical importance of adults is justified by the work of Ekou Nuama (2006, paragraph 32). The author shows that age can have positive externalities in the sense that older producers may be efficient because of their experience. In addition, supply to the Boundiali market depends on cereals, oilseeds and tubers from rural areas. However, studies by Sophie Pulcherie TAPE (2023: 4) show that the foodstuffs available in Ayamé are dominated by plantain, cassava and taro from the surrounding countryside. The transport of these rural food products from the production basins to the physical trade centre at Boundiali is carried out by various means of transport. Only two- or three-wheeled vehicles are in high demand. The use of two- or three-wheeled vehicles is due partly to the low volume of production and partly to the distance of the roads. This reality of the study area is confirmed by the work of Djibril KONATE et al (2021: 3128), who argue that the use of lorry capacity depends on the size of the products to be transported and the distance from the product collection area. However, the means of transport used in Boundiali is different from that used in Téra in Niger (Djibril KONATE et al., 2021, op cit). Pack animals are the means of mobility in the area. To reconcile all these positions, Von Thünen's theory states that distance justifies transport costs. For this However, there are some places that are better than others for a given activity. The further or more distant the food production areas, the higher the cost of transporting foodstuffs. Access to the Boundiali market for local food products provides an opportunity for producers, traders and the town council to generate financial resources. There are two channels for marketing food products: the short or direct channel and the long or indirect channel. Producers sell their goods either directly to consumers or to traders. This method of selling local food products is in line with the findings of Sophie Pulchérie TAPE et al (2020: 19). These authors indicate that 10% of producers sell their products to consumers, then 47% to traders and finally 43% to both consumers and traders in the town of Bouaké. As a result, the collection of taxes through producer-sellers has bolstered the municipal budget. This economic improvement of the local authority through taxes and fees is also justified by the work of ALE Agbachi Georges (2020: 8), who argues that revenue represents a considerable financial windfall for the communal budget as part of its territorial development. This author, in line with MBELLA MBONG Rostant and MAWO Vigenie Yoland (2023: 189), shows the effectiveness of market management by the local authority of Ouaké in north-east Benin. He argues that the local authority has collected an average of between 22 and 40 million CFA francs since 2012 in revenue from taxes and fees for occupying the market.The Boundiali market is where foodstuffs and manufactured goods are sold. This is borne out by the work of WILHEM (1997: 20), who maintains that the urban market is the place where the urban population obtains supplies of manufactured goods, foodstuffs and basic necessities. In addition, this physical space for commercial exchange receives users from a variety of backgrounds. It is the melting pot of a cosmopolitan social stratum. Both men and women are involved in trading in the area. In this respect, there is a numerical superiority of female players to the detriment of so-called male players. The arithmetical subjugation of men in favour of women is also seen in the work of Nadège Ahou Kouadio Konan (2021: 8) in the large market of Treichville and the market forum of Adjamé in Abidjan. This author confirms that women are in the majority in the marketing of foodstuffs, with a rate of 71.84%. This position is supported by Guillaume Hensel FONGANG FOUEPE et al, (2019, op cit). In fact, according to them, the plantain banana trade in the West Cameroon region is run by women, 84% of whom are women. The cultural mix created by the meeting of sellers, buyers and consumers attracted our attention. The study on the origins of the various players favours nationals. In other words, the foodstuffs trade in the Boundiali market is concentrated in the hands of Ivorians, even though there is substantial participation by non-nationals, namely Malians and Mauritanians. The unbridled configuration of actors in the marketing chain in the transaction areas of the study area is in perfect symbiosis with the results produced by Patrice MOUNDZA and Robert Edmond ZIAVOULA (2006: 227), which show that the majority of traders and sellers in the markets of the city of Brazzaville are of Congolese nationality. Also, studies subsequent to the previous ones relating to education have challenged DAKOURI Guissa Francis and KOULAÏ Armand (2015: 5). For these authors, the dominant level of education in the markets of Abobo-Centre in Abidjan is that of the illiterate, with a rate of 55%, compared with rates of 24%, 19% and 2% respectively for primary, secondary and higher education. The positions of the above-mentioned authors thus consolidate the results of our work, since the majority of traders in our study area are illiterate. In addition, the Boundiali trading area is busy every day. On Saturdays, however, there are large flows of sellers, buyers and goods from various locations. This lived reality is perfectly similar to that developed by ALE Agbachi Georges (2020: 1). For him, the Kassoua-Allah market, which is held every Tuesday and located in the north-east of Benin close to the Togolese border, represents a dynamic hub of activity and trade. The overcrowding and overuse of this commercial edifice, especially on the flagship day, tarnish the quality of environmental hygiene. To this end, the perimeter used for trading products, particularly foodstuffs, is identified by its unsuitable environment. It is a receptacle for all kinds of waste and wastewater. On closer examination, the areas where foodstuffs are displayed are more polluting than those where manufactured products are sold. What's more, the alleyways near the markets are much sought-after on busy days. This reasoning is borne out by various scientific studies and writings. In a pragmatic way, BELLEFLEUR Steve (2019: 54) explains that the spatial aspect of markets and affirms that the cramped nature of markets and the numerical insufficiency of spaces lead to the occupation of nearby arteries. In addition to this form of pollution, Patrice Moundza and Robert Edmond Ziavoula (2006, op. cit.) show that rubbish is strewn around all the markets in Brazzaville. WHO (2001: 32) adds that goods are sold in unsanitary areas.

CONCLUSION

Food production requires the use of gifts, inheritance, tenancy and sharecropping. However, the inheritance method is emerging. Land tenure has an impact on the area sown. Maize (36.20%), rice (21.27%), millet (1.84%), sorghum (0.61%), beans (0.41%), groundnuts (35.38%) and yams (4.29%) make up the areas sown. Other crops include maize (37.85%), groundnuts (37.38%), rice (22.33%), yams (1.87%), millet (0.36%), sorghum (0.15%) and beans (0.02%). Chicken (36.19%), goat (8.25%), millet (0.25%), sorghum (0.15%) and beans (0.02%) were the other main crops. Mutton (23.92%), beef (27.20%), pork (1.27%) and guinea fowl (3.17%) are the main food animal products in the study block. These foodstuffs reach the Boundiali market by means of transport, with two- or three-wheeled vehicles (48.67%) being the most popular. The sale of foodstuffs opens up a way for the local authority to reap the financial resources that are essential for communal development. However, the marketing of food products is leading to a deterioration in the environment of the Boundiali market. For this reason, the town council should encourage food producers to form a professional agricultural organisation, distribute agricultural inputs free of charge, allocate subsidies, maintain roads and ensure that the market is kept clean.

REFERENCES

ALE Agbachi Georges (2020), Marchés ruraux et dynamique spatiale au nord Bénin : l'exemple du marché de Kassoua- Allah à Ouaké, Annale de l'Université de Moundou, Série A-FLASH Vol 7 (:), 12p.

Anne-Marie COTTEN (1968) Les villes de Côte d'Ivoire. Une méthode d'approche par l'étude des équipements tertiaires. Bulletins de l'Association de Géographes Français 45 (366), pp 223-238

BELLEFLEUR Steve, 2019: Proposition des stratégies de gestion des déchets générés par les marchés Fer, Relais et Jeudi dans la ville des, Université Notre-Dame d'Haïti (UNDH), Faculté d'agronomie, Mémoire de licence universitaire, 81p.

Céline Yolande KOFFIE-BIKPO, Akoua Assunta ADAYE (2014), Agriculture commerciale à Abidjan : le cas des cultures maraîchères. In for 2014/4 (N°224), pp 141-149

CHALEARD Jean Louis (1996), Marchés et Vivrier marchand en Afrique occidentale : le cas de la Côte d'Ivoire, Historians and geographers, p111-122

DAKOURI Guissa Desmos Francis, KOULAÏ Armand (2015): Commercialisation des produits vivriers et la dégradation de l'environnement dans les marchés d'Abobo-centre (Abidjan-Côte d'Ivoire), Revue de Géographie Tropicale et d'Environnement, n° 2, pp 66-76.

Djibril KONATE, Lacina FOFANA, Soualiho SOUMAHORO (2021), " Impact socio-économique et environnemental du bitumage d'une voie express : Exemple de l'axe Korhogo-Karakoro ", Journal Africain de Communication Scientifique et Technologique n°100, pp 3121-3134

DUFUMIER. Marc (1999), La prise en compte des risques dans la définition des politiques de développement agricole. In le risque en agriculture, Edition ORSTOM, Paris, pp547-560

Ekou Nuama (2006), Mesure de l'efficacité technique des agricultrices vivrières en Côte-d'Ivoire. Economie Rurale, n°296, pp39-53

Guillaume Hensel FONGANG FOUEPE, Achile BIKOI, Denis Pompidou FOLEFACK, (2019), Socioeconomic analysis of the plantain marketing system in the West Cameroon region, Int.J. Biol. Chem. Sci. 13 (4) : pp2259-2274

HOUNGBO Nounagnon Emile (2015), " Relation campagne et ville : Deux réalités complémentaires et interdépendantes ", In : AGRIDAPE, Relations ville-campagne, Num 31, vol 2, pp.13-14.

Jacqueline PELTRE WURTZ, STECK Benjamin (1991), Les charrues de la Bagoué. Gestion paysanne d'une operation cotonnière en Côte-d'Ivoire. Paris, Éditions de l'ORSTOM, 303 p.

KONAN.K.H., KRA.K.J., GOGOUA.G.E. (2016), "Les défis de l'approvisionnement de la ville de Korhogo en produits vivriers ", Journal des Sciences Sociales N° SPECIAL " Variations subsaharienne ", pp 47-62

KOUADIO D.B. (2004), Les micro-barrages en terre réalisés par les projets de développement de l'élevage en Côte d'Ivoire. Projet de développement de l'élevage-phase II BAD-CI, 44 p

Kouadio Konan N.A. (2021), Pratiques Sociales et Déficit d'Hygiène des Aliments au Sein du Grand de Treichville et le Forum des marchés d'Adjamé (Côte d'Ivoire). European Scientific Journal, ESJ, 17 (9), pp71-88

Lehou Franck Cyril TAPE BIDI, BELI Didier YAO, Céline Yolande KOFFIE-BIKPO (2018), Le commerce du riz imported into Abidjan (Côte d'Ivoire). EDUCI Revue de Géographie Tropicale et d'Environnement, n°2, p 127-138

Mathieu Juliot Mpabe Bodjongo, Fondo Sikod, (2017), Accès aux marches urbains et variation des revenus des agricultures ruraux du secteur informel au Cameroun. Revue D'Economie Régionale et Urbaine vol (2), pp357-378

MBELLA MBONG Rostant, MAWO Vigenie Yoland (2023), Des marchés ruraux au cœur du développement local : Cas des produits agricoles dans l'arrondissement de Mélong. Revue Espace Géographique et Société Marocaine, Numéro 67, pp 181-204

ONU-Habitat (2012), Côte d'Ivoire: Profil urbain de Boundiali, United Nations Human Settlements Programme, ISBN Number: (Volume) 978-92-1-132472-3, 30p

Patience MPANZU BALOMBA, Philipe LEBAILLY, Charles KINKELA SAVY (2011), Les conditions de productions et de mise sur le marché des produits vivriers paysans dans la Province du Bas-Congo (R.D.Congo). Les Cahiers de l'Association Tiers-Monde, n° 26, p 1.

Patrice MOUNDZA, Robert Edmond ZIAVOULA, 2006: Les marchés : Lieux de Ravitailment et alternatives à l'inactivité, Brazzaville, une ville à reconstruire, Paris, Karthala, 225-234.

Patricio Mendez del Villar, Jean-Martin Bauer (2013), Le riz en Afrique de l'ouest : dynamiques, politiques et perspectives. Cahiers Agricultures 22 (5), pp 336-344

René Kinimo YABILE (1986), Autosuffisance alimentaire en Côte d'Ivoire: Paradoxe ou Réalité socio-économique. Economie rurale 175, PP.44-49

SCHWARTZ Daniel (1995), L'échantillonnage : du prélèvement à l'analyse. Edition ORSTOM, 209 p.

SILUE Pébanagnanan David (2012), Impact socio-spatial des retenues d'eau dans le Nord de la Côte d'Ivoire : Cas de la région des Savanes. PhD thesis in Geography, Université Félix Houphouët Boigny d'Abidjan, 319 p

Sophie Pulchérie TAPE (2023), Circuit de commercialisation des produits vivriers dans le département d'Ayamé (Sud- Est de la Côte d'Ivoire), Université d'Abomey-Calavi. Journal de Géographie Rurale Appliquée et Développement N° 001, vol 4, pp. 53-66

Sophie Pulchérie TAPE, Médé Roger DINDJI, Kouakou Valère KOUASSI

(2020), Approvisionnement de la ville de Bouaké (Centre de la Côte d'Ivoire) en produits vivriers, University of Lomé. Revue de Géographie du Lardymes N° 25,pp. 95-107.

WHO, 2001: Health markets: a guide to hygienic conditions in food markets, WHO, 46p.

WILHEM Laurence, 1997: Les circuits d'approvisionnement alimentaires des villes et le fonctionnement des marchés (Urban food supply channels and market operations).in Africa and Madagascar. Communication, "Food in cities" FAO, 60p.

TABLE OF CONTENTS

INTRODUCTION 3

MATERIALS AND METHODS 5

RESULTS 10

DISCUSSION 56

CONCLUSION 60

REFERENCES 61

MIX
Papier aus verantwortungsvollen Quellen
Paper from responsible sources
FSC® C105338

Printed by Books on Demand GmbH, Norderstedt / Germany